Busy Bugs

BEETLES

By Bray Jacobson

Please visit our website, www.garethstevens.com. For a free color catalog of all our high-quality books, call toll free 1-800-542-2595 or fax 1-877-542-2596.

Library of Congress Cataloging-in-Publication Data
Names: Jacobson, Bray, author.
Title: Beetles / Bray Jacobson.
Description: New York : Gareth Stevens Publishing, [2022] | Series: Busy bugs | Includes index.
Identifiers: LCCN 2020006185 | ISBN 9781538263259 (paperback) | ISBN 9781538263266 (6 Pack) | ISBN 9781538263273 (library binding) | ISBN 9781538263280 (ebook)
Subjects: LCSH: Beetles–Juvenile literature.
Classification: LCC QL576.2 .J34 2022 | DDC 595.76–dc23
LC record available at https://lccn.loc.gov/2020006185

First Edition

Published in 2022 by
Gareth Stevens Publishing
111 East 14th Street, Suite 349
New York, NY 10003

Editor: Kristen Nelson
Designer: Katelyn E. Reynolds

Photo credits: Cover, p. 1 arlindo71/ E+ / Getty Images Plus; p. 5 Jake Jung/Moment/Getty Images; p. 7 phototrip/ iStock / Getty Images Plus; pp. 9, 24 (wings) Savas Sener/ 500px Prime/Getty Images; p. 11 Paul Starosta/Stone/Getty Images; pp. 13, 24 (eggs and grub) Christiana Fletcher/ iStock / Getty Images Plus; p. 15 YinYang/E+/Getty Images; p 17 mikroman6/Moment Open/Getty Images; p. 19 sandra standbridge/ Moment/Getty Images; p. 21 Siegfried Grassegger/Getty Images; p. 23 Gail Shumway/ Photographer's Choice / Getty Images Plus.

Printed in the United States of America

CPSIA compliance information: Batch #CSGS22: For further information contact Gareth Stevens, New York, New York at 1-800-542-2595.

Contents

Beetles live all over the world.

There are many kinds. They can live on land or in water.

All have six legs.
Most have wings.

Not all kinds fly.

They lay eggs.
Babies are grubs.
Their bodies grow
and change.

Most eat plants.

Ladybugs eat other bugs!

They can be pests.
They may kill plants!

Some kinds come out
at night.

Fireflies glow!
They live in
warm places.

Words to Know

eggs

grub

wings

Index